RAINTREE
SCIENCE ADVENTURES

UNDER THE SEA

Helen H. Carey and Judith E. Greenberg
Illustrated by Andrea Z. Tachiera

Raintree Publishers
Milwaukee

For Jack — **H.C.**
To Jonathan, Carolyn, and Hilary — **J.G.**

Editorial
Barbara J. Behm, Project Editor
Judith Smart, Editor-in-Chief

Art/Production
Suzanne Beck, Art Director
Kathleen A. Hartnett, Designer
Carole Kramer, Designer
Andrew Rupniewski, Production Manager
Eileen Rickey, Typesetter

Reviewed for accuracy by:
Gretchen M. Alexander, Executive Director
West 40 Educational Service Center Number 5
Northlake, Illinois

Scott R. Welty, Instructor
Maine East High School
Park Ridge, Illinois

Library of Congress Number: 89-78283

1 2 3 4 5 6 7 8 9 94 93 92 91 90

Library of Congress Cataloging-in-Publication Data

Carey, Helen H.
Under the sea / by Helen H. Carey and Judith E. Greenberg; illustrated by Andrea Z. Tachiera.
(Raintree science adventures)
Summary: Describes what it is like to search for shipwrecks from on board a research ship and to live and study in an underwater science laboratory. 1. Underwater exploration—Juvenile literature. 2. Manned undersea research stations—Juvenile literature. [1. Underwater exploration.] I. Greenberg, Judith E. II. Tachiera, Andrea Z., ill. III. Title. IV. Series.

GC65.G714 1990 627.7 89-78283
ISBN 0-8172-3755-0 (lib. bdg.)

Before You Begin

This book takes *you* on an underwater adventure! First, you will explore the surface of the ocean, or sea. Then you will go deeper underwater to explore and to live for a time. You will become part of a new world, one that is almost as strange as another planet.

You will also do two experiments that will help you understand water pressure and the darkness of the sea floor.

Knowing the words below will help you in your adventure.

apparatus equipment designed for a special use
coral a substance made up of the outer skeletons of tiny sea animals
food chain a group of plants and/or animals in which each member feeds on the member below it and is then eaten by the member above it
habitat the place where a plant or animal normally lives and grows
hydrothermal relating to hot water
marine relating to the sea
microscope an instrument that can enlarge the appearance of tiny objects so that they can be seen
snorkel a tube that lets a person put his or her face under the water and still breathe air from above water
submarine a boat that can operate underwater
submersible something, such as a boat or a robot, that can go underwater

Now turn the page, and begin your science adventure!

A Mystery Map

You are on the beach, getting ready to dig a hole in the sand. Suddenly, you notice a crumpled piece of paper near your shovel. You unfold the paper and look at it. It is a map. There is an *X* in the middle. You know that pirates used to mark with an *X* the spot where they buried treasure. You must have found a treasure map!

You start to dig in the sand. If you keep digging, you might find the treasure. There might be gold coins and jewels! You dig and dig until the hole is so deep you can stand in it up to your elbows. You don't find any treasure or even any clues in the hole.

Holding the map close to your body so no one else can see it, you look at the map again. You realize that the *X* isn't on the beach. It is in the ocean.

You decide to go looking for treasure in the ocean. You roll the map in your towel and shove it under your beach blanket. At the edge of the ocean, you put on your life jacket, your **snorkeling** mask, and your flippers. You tell your big sister, Karen, that you are going into the water for a while. You wave to the lifeguard. She nods back to let you know that she will watch you in the water.

You enter the water. The only sound you hear is the air going in and out of your snorkel. The water is bright and clear. You reach out to touch an angelfish, but it darts quickly away.

You watch a big fish try to eat a little butterfly fish. On the little butterfly fish's tail, there is a black spot that looks like an eye. The bigger fish moves in on the "eye." Instead of a meal, however, the bigger fish gets a mouthful of water as the little fish darts away. Many fish in the sea have markings, colors, or shapes that help them escape or hide from their enemies.

You swim for a long time, but you do not see any hidden treasure. You are tired but not ready to give up your search. You go back to the beach to look at the map again.

The *X* is farther out in the ocean than you first thought. You will need more than your snorkeling equipment to find the pirate treasure. You will need scuba-diving equipment. *Scuba* is a short word that means *s*elf-*c*ontained *u*nderwater *b*reathing **apparatus.** Scuba divers breathe air stored in a tank on their back.

Your friend Bud is a scuba diver. You visit his diving equipment shop on the beach.

You tell Bud about the treasure map, but he won't lend the scuba equipment to you. He tells you that people must be trained before it is safe for them to go scuba diving.

Bud looks closely at the map and shakes his head. "This isn't a real pirate map. I don't know what the *X* marks, but you're right about it being in the ocean." He hands you back your map.

Outside the shop, you sit down and sift some sand through your fingers while you think. Why would anyone make an *X* on a map of the ocean?

Bud comes out and says, "You should talk to Dr. Wong. She knows all about finding treasure in this part of the ocean."

Underwater Equipment

You find out where Dr. Wong's office is, and Karen drives you there. You know you are outside the right office because the door is propped open with a cannonball. A cannon takes up most of the office inside. A sign on the cannon says that it is from a Spanish pirate ship.

A voice from behind the cannon calls out, "Ahoy, pirate! Make them walk the plank!" A parrot flies from behind the cannon and through a door leading into a large workroom. The parrot perches on a woman's shoulder.

The woman standing there smiles at you. "Don't mind Captain Kidd. The silly bird thinks everybody's a pirate! I'm Dr. Wong. Can I help you?" she asks.

You hand her the map. "I found this on the beach. I want to know if it's a treasure map," you say.

"You found my map!" exclaims Dr. Wong. "I lost it yesterday on the beach. The *X* marks the place in the ocean where we're searching for a sunken ship."

"Are you looking for a pirate ship with gold and jewels on it?" you ask.

"Not exactly," Dr. Wong answers. "I am a **marine** archaeologist. I study old objects that are found in the sea. The things I find were once the ships' cargo, or were worn by sailors, or are parts of the sunken ships themselves. Sometimes gold coins and jewels can be found, but not often."

"How far down into the ocean can you go? How long can you stay down there? Will you take me with you sometime?" you ask, all in one breath.

Dr. Wong answers, "Whoa. Slow down. The answer to your first question is found on this chart." She leads you to a big blue chart. The chart shows the various depths in the ocean that people explore. It shows vehicles and other pieces of equipment that are used underwater. Scientists have given them special names.

"People have different reasons for entering the undersea world. Some people dive to find scientific information on minerals, fish, and sea plants. Some go down to build or repair underwater bridge supports, pipes, and cables. Divers also help recover sunken ships or treasure," explains Dr. Wong.

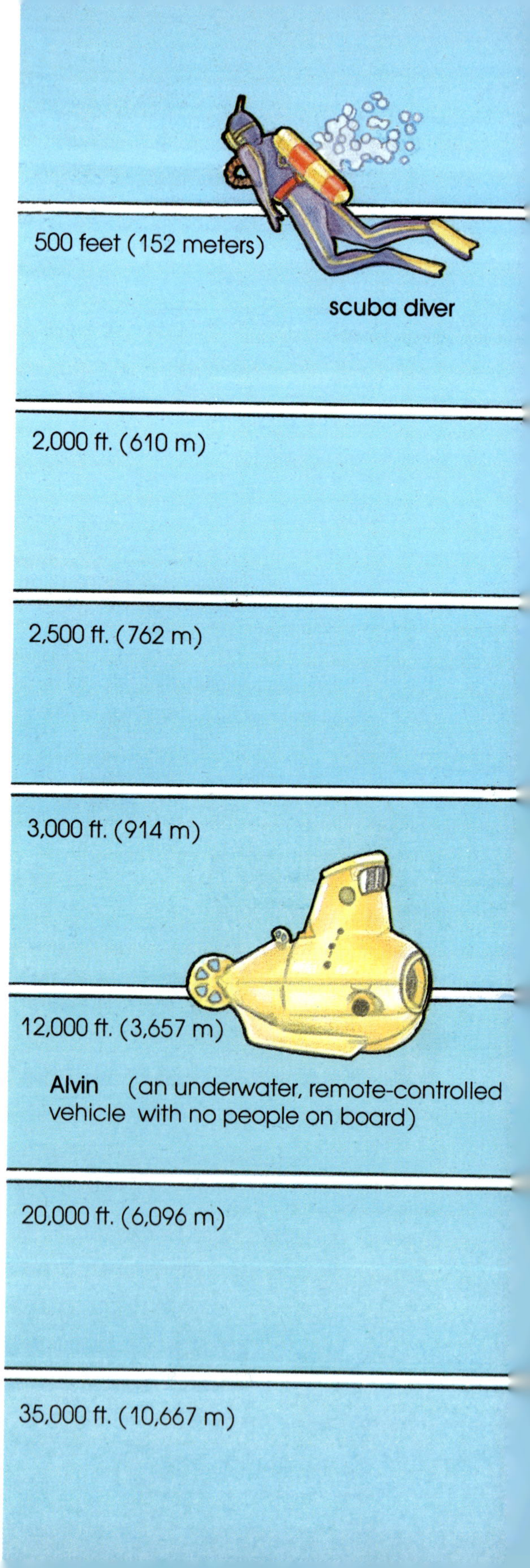

JIM (a diving suit made of rigid metal sections that pivot. It protects the diver from the cold and pressure of deep water. JIM is named after Jim Jarrett, who invented the suit.)

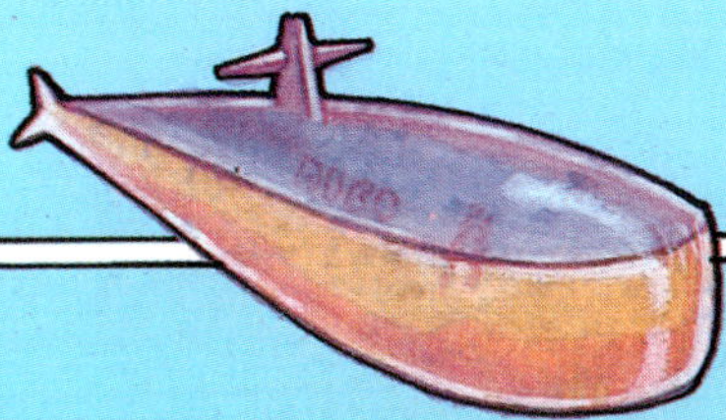

submarine (a ship that can operate on the surface or underwater)

bathysphere (a steel diving sphere for deep-sea observation)

JASON (a vehicle that photographs and retrieves underwater objects and plant and animal life through remote control)

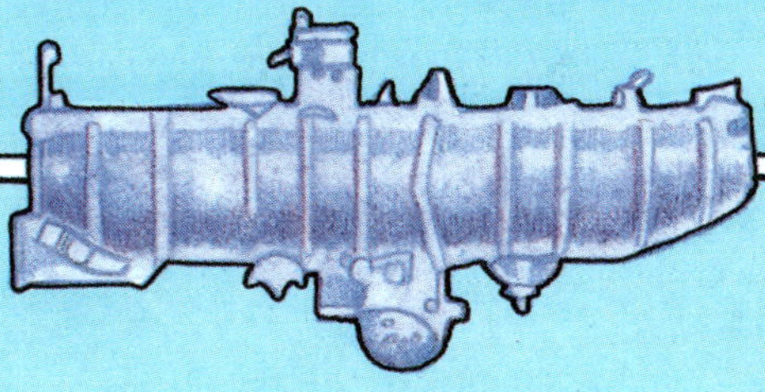

Trieste (a bathyscaphe, or underwater ship, used for deep-sea exploration that has a rounded cabin attached to its underside)

A voice calls out, "JIM, JIM!"

"Who's JIM?" you ask, thinking you are going to meet another archaeologist.

Dr. Wong laughs. "Captain Kidd wants you to come into the next room. That's where JIM is," she says.

Captain Kidd perches on top of something that looks like a diving suit made out of metal sections put together.

Dr. Wong explains, "JIM is the name of this diving suit. It is made of lightweight metal. The person wearing it can walk on the sea floor. The diver breathes air from a tank inside the suit. The diver's hands are also inside the suit. He or she moves levers inside JIM to operate claws found on the outside of the suit. These claws open and close to grip things and pick things up."

Dr. Wong goes to answer the telephone. In the corner, you recognize a big round object that is on Dr. Wong's chart. It is a bathysphere. The door on it is hanging open.

You climb into the bathysphere, and the door closes behind you. You push on the door, but it doesn't open. You feel scared. Dr. Wong can't hear you calling for help. You see a flash of feathers as Captain Kidd flies around the outside of the bathysphere. Soon after, Dr. Wong opens the door and lets you out.

"Captain Kidd saved you. You would have needed air soon," she says. You gasp a little and relax. You ask Dr. Wong how people can breathe underwater in the bathysphere.

"The bathysphere has an air supply available when the sphere is underwater. The bathysphere also protects divers from very cold water and the weight or pressure of the water," she answers.

You frown. You don't understand water pressure.

Dr. Wong leads you to a sink. She punches three holes of the same size in a milk carton. She puts a piece of tape over each hole and fills the carton with water. She places the carton beside the sink and asks you to take the tape off the top hole. Water pours out.

You mark the place where the stream of water lands in the sink. You replace the tape over the hole and refill the carton with water. You do the same thing with each of the two other holes.

"The stream from the bottom hole went the longest distance," Dr. Wong says. "The bottom hole had more water above it than at the other two holes. The pressure of all the water at the bottom hole pushed that stream the greatest distance. The deeper you go into the ocean, the more pressure there is on your body because of the amount of water above you."

You wonder what the deep part of the ocean looks like.

Dr. Wong gives you a large, green plastic soda bottle and a flashlight. She says, "Very carefully, use scissors to cut the bottle into several pieces. The pieces should be large enough to cover the lens of the flashlight.

"When it's dark, go into your room and turn off all the lights. Turn on the flashlight. Cover the flashlight with the pieces of plastic, one piece at a time. Although the ocean isn't green, each piece of the green plastic that you add will block more and more of the light. This is what happens to light as you go deeper and deeper into the ocean. Stop when it becomes too dark to see the light from the flashlight. Then try to answer the question of how the plastic and flashlight are like the ocean and sunlight," Dr. Wong says.

(Write your guess on a piece of paper.)

It is getting late, and Dr. Wong is kind enough to drive you back to your house. When you return to visit Dr. Wong the next morning, there is another scientist with her. You shake hands with Dr. Wagner, who is a marine biologist. He tells you, "I study all the different kinds of animals and plants in the sea. Right now I am studying creatures that live in the deepest part of the ocean. These creatures live in the dark around openings in the sea floor."

Your smile stretches from ear to ear. The experiment with the flashlight helps you understand what Dr. Wagner means about the darkness on the floor of the sea. You show your notebook to both scientists. They look at your notes from last night.

Room is dark.
1 piece of green plastic over flashlight: I see everything in my room.
2 pieces over flashlight: I still see a lot.
3 pieces: It's getting darker.
4 pieces: It's not easy to see.
5 piec It's about the same.
6 pieces: It's about the same.
7 pieces: There's just a little bit of light.
8 pieces: I can only see my hand.
9 pieces: Now it's dark.

"It gets darker in the ocean as you go deeper. At the bottom of the ocean, the water blocks out all the sunlight from above. No light can get through," says Dr. Wong.

Dr. Wagner explains that plants and animals that live in the dark on the bottom of the ocean stay near openings in the sea floor. These openings are called **hydrothermal** vents. He tells you that the plants and animals live off of chemicals coming out of the vents.

You want to know how the hydrothermal vents were discovered. "If the vents are at the bottom of the ocean, wouldn't it be too dark for people to see them?" you ask.

Dr. Wagner says, "People travel to the ocean floor in machines called **submersibles** that have lights attached to them. Submersibles can go very deep. Sometimes there are one or more people in a submersible—for example, in a **submarine**. Sometimes the submersible is a robot without any people in it.

"Look, here's a model of the kind of underwater vehicle that we use. It's a ROV," says Dr. Wagner, pointing to a model on the table. "*ROV* stands for *r*emotely *o*perated *v*ehicle. It is a little like a toy remote-control car. ROV, however, is bigger and goes under the water. The ROV's claws gather plants and animals in nets for us to study."

The Searchers

Dr. Wong has invited you to go aboard the big research ship where she is working this summer. You meet many interesting people on the research ship.

An oceanographer is a scientist who studies the ocean—for example, the water, tides, and currents. Oceanographers spend much of their time working in or near the ocean, but they also work in laboratories on land.

Crew members help lower the ROV into the water. The crew runs the ship, cooks the food, and assists the scientists.

A marine geologist is a scientist who studies the solid mass that makes up the sea floor. Marine geologists make maps of the land under the sea and look for valuable minerals. Some marine geologists help energy companies find places in the ocean to drill for oil and natural gas.

A marine biologist is a scientist who studies all the living things in the ocean. For example, the marine biologist might watch for changes in ocean plants and animals to learn what effects pollution has upon them.

Skilled divers make underwater repairs, take photographs, search for wrecks, and explore the sea. Some divers get their air from a hose connected to a ship. These divers can stay underwater longer but cannot move as freely as the scuba diver.

A marine archaeologist searches the sea for things used by people in the past. The marine archaeologist looks at the wrecks of sunken ships to help learn how people lived a long time ago.

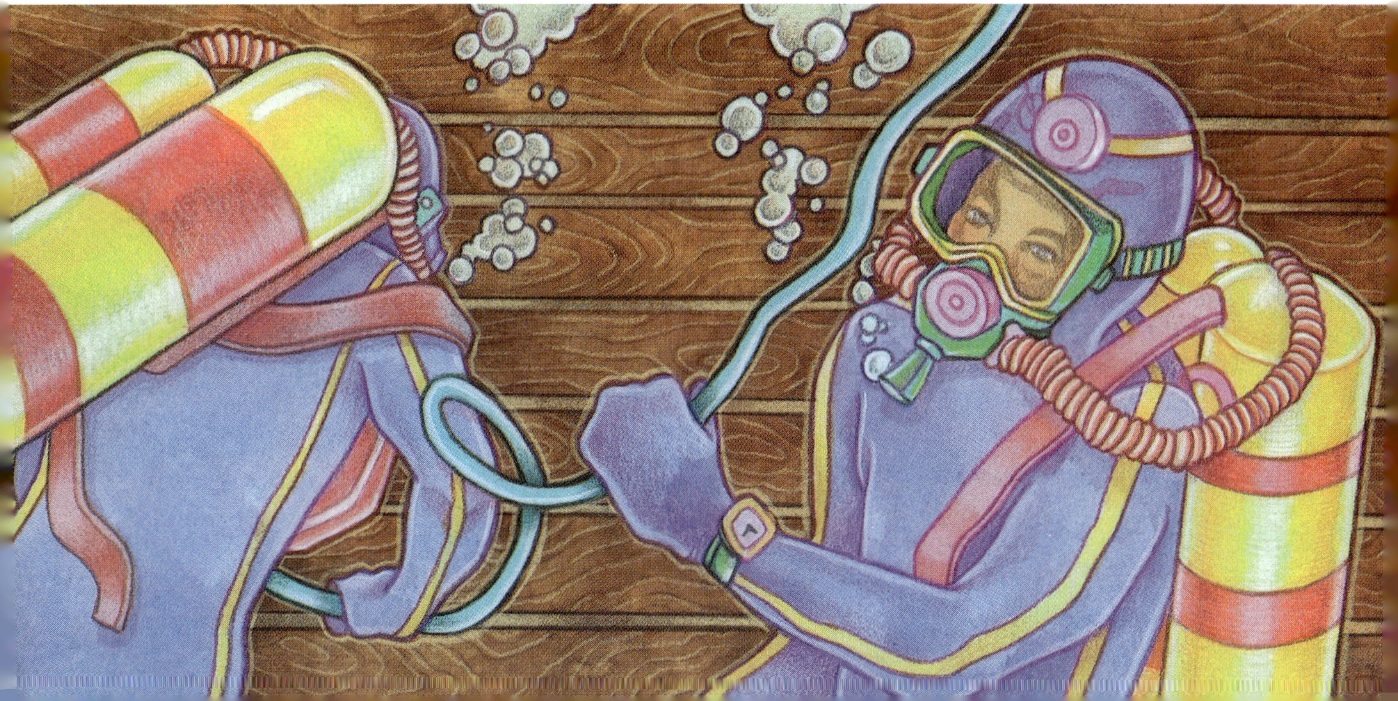

As the youngest person on board, you are asked to think of a name for the ROV. The crew likes the name you choose. They paint the name *Rover* on the vehicle.

Later on, the ROV is lowered into the water. Rover's video cameras and big lights sweep over the ocean floor. On the research ship, you watch a monitor, which looks like a television screen. It shows you what Rover is locating.

Rover travels to the area where a sunken ship has settled. Dr. Wong says, "Rover is right on the *X* spot that is on the map. Watch the monitor carefully. We hope that Rover's camera will find pieces of the ship or the cargo."

"I think I see an old sword!" you shout. "Can Rover pick it up?"

"You've got sharp eyes!" says Dr. Wong, patting you on the back. "Rover isn't going to pick up anything today, however. First, we have to make a diagram that shows the position of everything on and near the sunken ship. After that is done, we can start bringing up the sword and other items.

"Everything we find is like a piece of a puzzle. The only way to learn about the people who sailed this ship is to keep the pieces in place until we've studied them.

"Don't be disappointed. Tomorrow we'll have another exciting day," Dr. Wong promises you.

In a Science Lab on the Ocean Floor

The next day finally arrives! You are very excited because you are going to explore under the sea. Dr. Wong's assistant, Greg, is also a scuba-diving instructor. He has been teaching you how to scuba dive. You squint in the bright sunlight as you put on your equipment. You are going with Greg to an underwater science laboratory called Neptune. Neptune is anchored 60 feet (18 meters) below the surface of the water.

You roll backward over the side of the ship. You hold Greg's hand as you swim deeper and deeper toward Neptune.

You see the lights of Neptune. You and Greg enter the floating underwater lab through an opening at the bottom of it. You take off your air tanks, climb up a ladder, and enter Neptune.

There are four men and two women marine biologists in the underwater laboratory. They are studying sea animals and plants in their natural ocean **habitat.** Greg tells you, "Underwater labs are very important to science. Before we had them, we could only study sea creatures by taking them out of the water for a few minutes at a time. Now we can watch the animals right where they live."

The air inside Neptune is warm and moist. You look out the large windows at a world where beautiful fish and other animals glide by. An octopus slides across a window and looks at you with its big eyes. You count its eight tentacles, or feelers, with the two rows of suction cups on each tentacle.

Two or three times a day, and sometimes at night, the marine biologists put on diving equipment and go outside Neptune. They are studying a **coral** reef.

You learn that coral reefs are underwater ridges that look like beautiful rock gardens. They are found near the shore in the warmer seas of the world. You also learn that coral is made up of the skeletons of tiny animals called polyps. Coral reefs are necessary to many fish. These fish lay their eggs, feed, and hide from enemies in the reefs.

You like learning about the coral reef. You ask Greg to tell you more about life in the ocean.

"We'll talk some more when I get back," says Greg. "Right now I'm going to collect plankton." Greg grabs a net and tells you he's going outside of the underwater laboratory.

"I'm going to use this net to collect plankton. Plankton is made up of tiny plants and animals that float in the water. You need a **microscope** to see the individual plants and animals," says Greg.

"What's so important about plankton?" you ask Greg.

He answers, "Plankton is an important part of the ocean's **food chain.** In this food chain, tiny plankton animals eat tiny plankton plants. Then the tiny animals are eaten by small fish who are eaten by bigger fish. Plants and animals that aren't eaten eventually die and sink to the bottom of the sea. Their bodies rot and break down into minerals. These minerals spread through the water and nourish the plankton plants."

When Greg comes back, he hands you a container of seawater. You look at a drop of the water through a microscope. "I can see all kinds of moving things. Some look like beautiful stars, and others look like hairy monsters," you say excitedly.

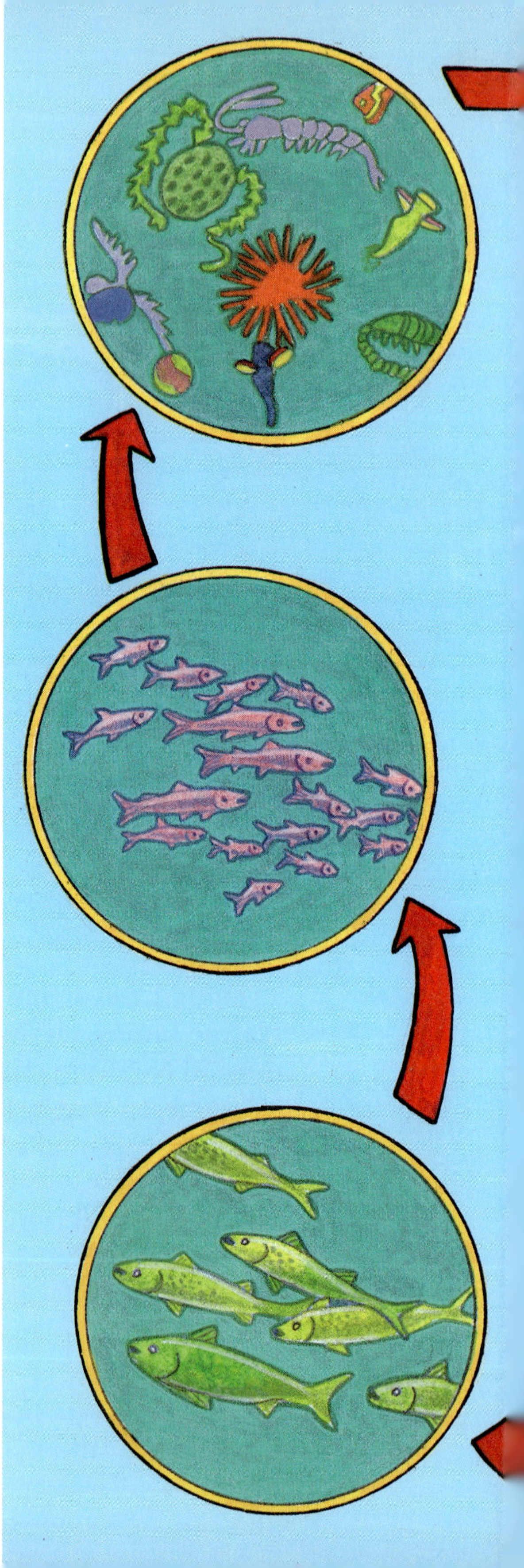

Greg says, "Those are plankton. Try to count the different kinds of plankton you see. Then draw pictures of them. Later, we will look them up in a book to see what kind they are. Some plankton live in clean water. Some live in polluted water. When we find out what kinds of plankton we have here, we can tell if this part of the sea is becoming polluted."

Before you know it, your undersea adventure comes to a close. During your swim to the surface, you decide that someday you want to be a scientist who works under the sea. You might be a marine biologist who studies water pollution or a marine archaeologist who finds all kinds of interesting things from days gone by. You have many other choices, too.

What kind of science adventure would you like to go on next?